WIND POWER

WIND POWER

By NORMAN F. SMITH
Illustrated by DON MADDEN

Coward, McCann & Geoghegan, Inc.
New York

Library of Congress Cataloging in Publication Data
Smith, Norman F.
Wind power.
SUMMARY: Describes why the wind blows and how it can
be harnassed through use of windmills to provide free,
non-polluting energy.
1. Wind power—Juvenile literature. [1. Wind power.
2. Winds] I. Madden, Don. II. Title.
TJ825.S69 621.4′5 80-28434
ISBN 0-698-30732-1 AACR1
First Printing
Printed in the United States of America

To my grandson
Hans Frederick Smith
N. F. S.

The wind blows most of the time at most places on the earth's surface. Sometimes the wind is a gentle breeze, bringing us cool air on a hot day. Sometimes it is strong enough to lift a kite into the air, to blow off your hat, or to blow an umbrella inside out. Sometimes the wind is a howling blizzard or a roaring hurricane with a force that can knock down trees and tear roofs off buildings.

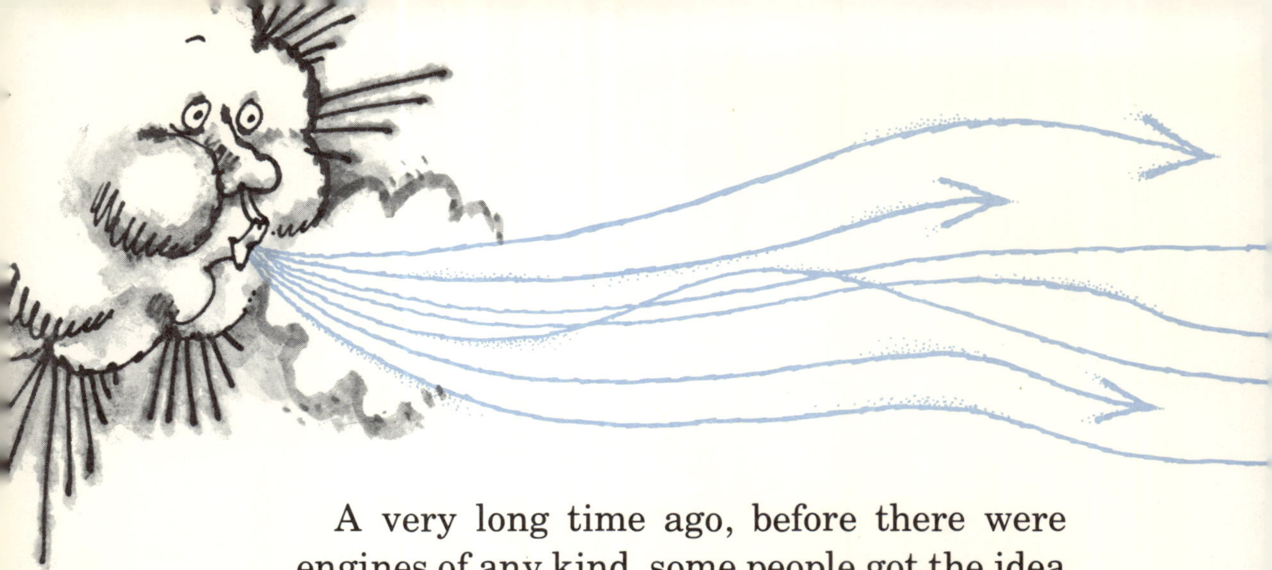

A very long time ago, before there were engines of any kind, some people got the idea of using the force of the wind to do useful work. They made a sail by sewing animal skins together and hung it on sticks to catch the wind. They found that using wind power to push their boats was much easier than using their own muscles to paddle the boats.

No one knows who invented the sail, or when they lived. Drawings and carvings found in Egypt show that the ancient Egyptians were using sails on their small boats back in the days when the pyramids were being built, over 5000 years ago. Later, the

ancient Phoenicians, Greeks and Romans built fine sailing ships. Some of them had oars so that the crew could row when the wind stopped blowing. By the time of Columbus, large sailing ships were being used to carry cargo and people over long distances.

Many hundreds of years after the sail was invented, some people thought of using the power of the wind to operate machinery. They built a windmill which had big sails or "vanes," probably made of cloth mounted on wooden frames. The wind pushed on the vanes and made the windmill spin like a giant pinwheel. The first windmill was probably built in one of the countries in the region now known as the Middle East. The force of the wind may have been used in this first windmill to pump water or to grind grain into flour.

The first windmill may have looked like this

Windmills have been used all over the world. Holland is famous for its windmills because thousands of them were once used there to pump water to keep the sea from flooding the country.

Windmills were also used in the United States in colonial days to grind grain into flour. The mill operator spread cloth sails on the wooden blades to catch the wind. Then he turned his windmill to point it into the wind. When the blades were turned by the force of the wind, they rotated shafts and gears that turned the grindstones. The work of grinding the grain was done by the wind. All the miller had to do was to pour in the corn or wheat and collect the ground-up flour.

Windmills were also used to saw lumber. Sawing logs into beams and boards in colonial days was very hard work. It was usually done by two men pulling a saw back and forth by muscle power. In the wind-driven sawmill the workmen only had to put the log in place and the power of the wind turned the machinery that moved the saw back and forth to cut the log into lumber.

Windmills have been used in various countries to grind chalk and spices, to polish stone, to crush sugar cane, and to make paper. You can still see small windmills today on farms and ranches, pumping water for farm animals where no electricity is available.

What makes the wind blow? The sun makes the wind blow because the sun warms the air in the earth's atmosphere more in some places than in others. When the air is warm in some parts of the world and cold elsewhere, air is forced to move from one place to another. We feel this moving air as *wind*.

You can see how the sun heats the earth by shining a flashlight on a globe. Pretend that the flashlight is the sun and the globe is the earth. When you shine the flashlight toward the center of the globe, the light shines very brightly on a small area. That's the way the sun's rays strike the earth near the equator. The strong sunlight falling on these regions makes them very warm.

When you shine the flashlight toward the edge of the globe, the light strikes a glancing blow and shines more weakly over a larger area. That's the way sunlight strikes regions near the north and south poles of the earth. The weak sunlight falling on these regions leaves them very cold.

Because the air in the earth's atmosphere is cold near the poles and warm near the equator, the air will move from one place to another. You can find out how cold air moves by opening the door of a refrigerator a few inches and holding your hand below, and above, the opening. Does the cold air flow upward or downward out of the refrigerator?

Hold your hand above a hot toaster or a warm stove (be careful not to touch the hot parts). Which way does the warm air move?

The warm air at the earth's equator, because it weighs less than the cool air around it, rises in the atmosphere like the air above a warm stove. The cool air at the earth's poles does just the opposite. Because this air is heavier than the air around it, it sinks downward in the atmosphere like the cold air flowing out of an open refrigerator. The cold air moving downward at the poles spreads over the earth toward the equator, while the warm air rising at the equator flows in the other direction, toward the poles.

If the earth were *not* rotating, the air movements between the poles and the equator of the earth would look like this:

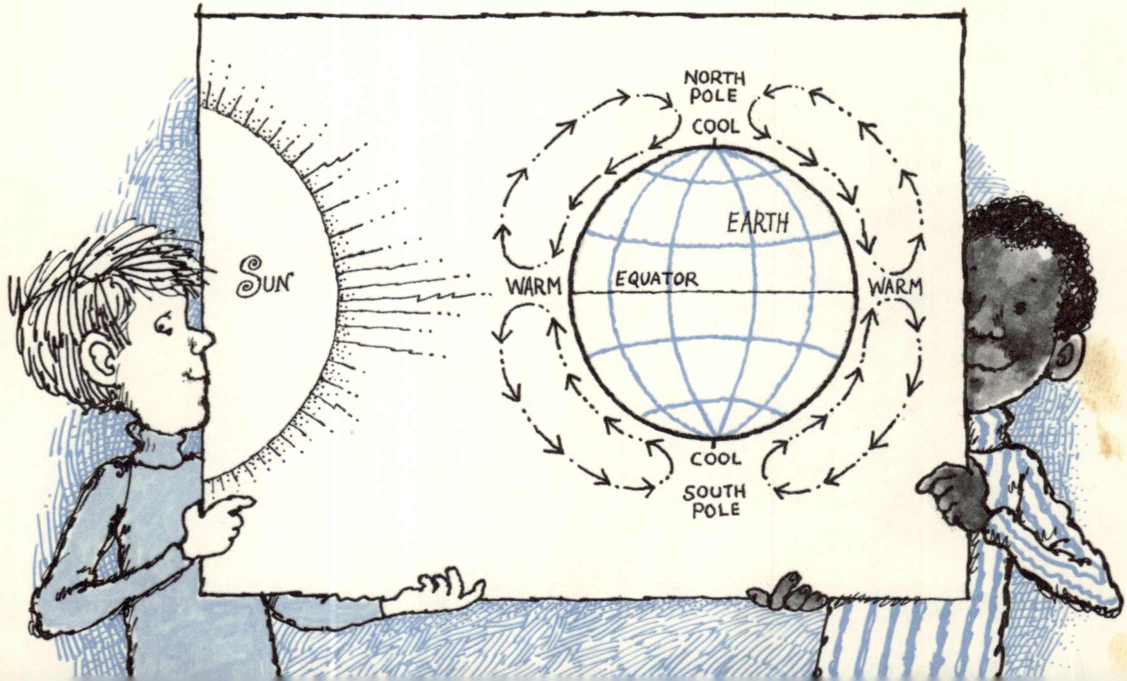

But the earth *does* rotate, and the rotation causes these wind patterns to break up into three whirling cells between the equator and each pole that look like this:

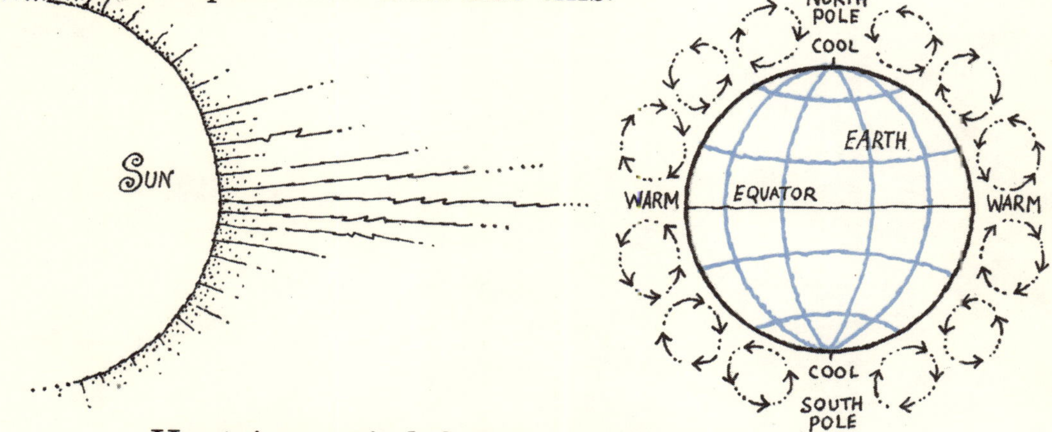

Heat is carried from one cell to the next from the warm equatorial region to the cooler regions toward the pole. The earth's rotation also twists the whirling cells so that the flow of the wind over the ground and over the ocean's surface looks like this:

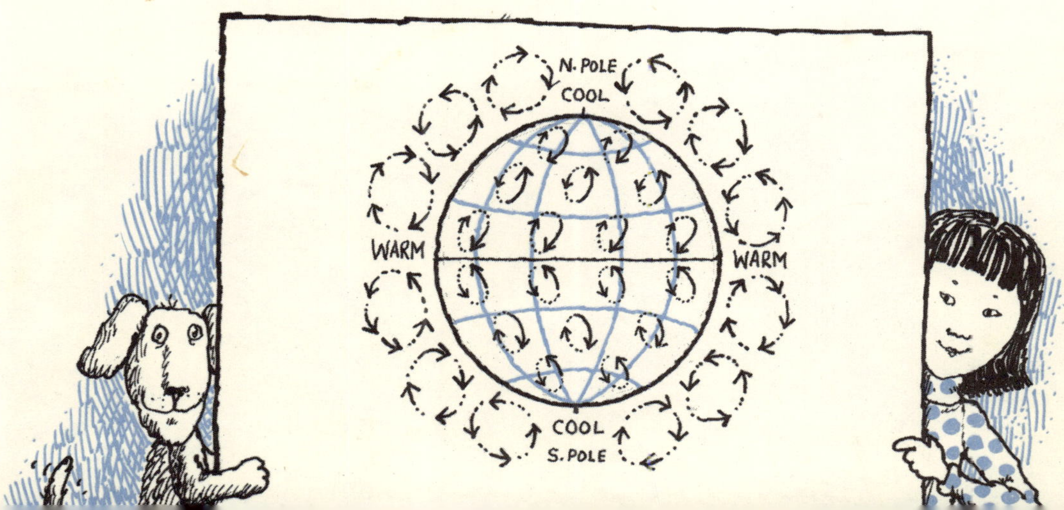

But the earth *does* rotate, and the rotation causes these wind patterns to break up into three whirling cells between the equator and each pole that look like this:

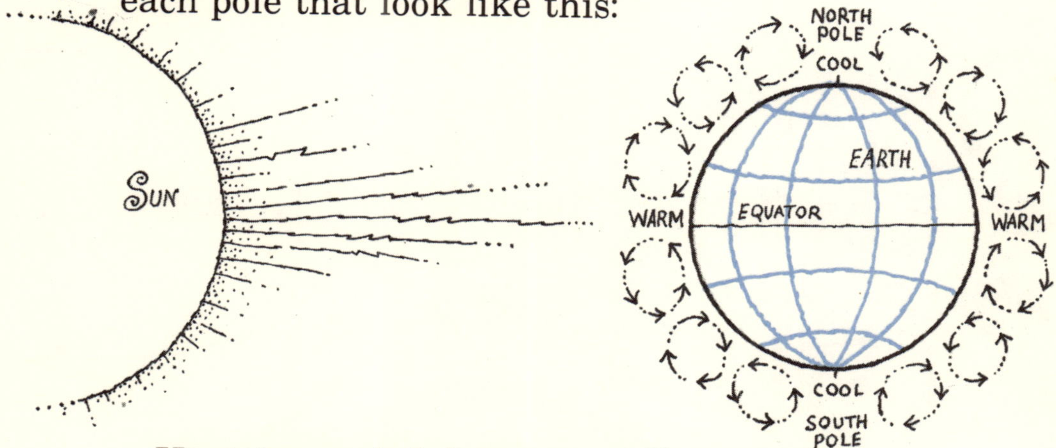

Heat is carried from one cell to the next from the warm equatorial region to the cooler regions toward the pole. The earth's rotation also twists the whirling cells so that the flow of the wind over the ground and over the ocean's surface looks like this:

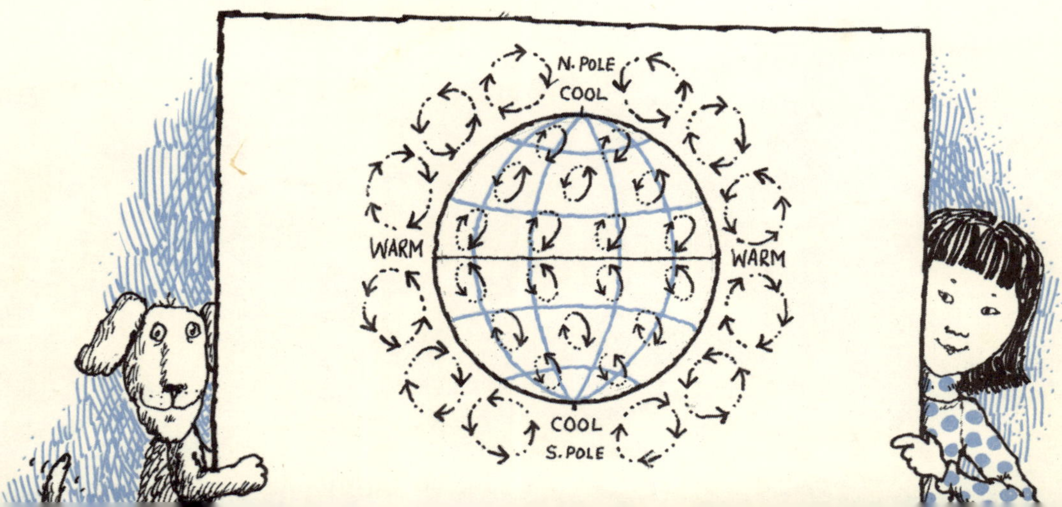

Like a giant pump, powered by the sun, the winds in the earth's atmosphere pump heat from the warm regions to the cool regions of the earth.

These winds flow across the earth's surface like great rivers of air. Sailors called some of these winds "trade winds" in the days of sailing ships. They learned that the trade winds blew most of the time, and used them to drive their ships from one seaport to another all over the world.

From this map, showing how the winds usually move over the Atlantic Ocean, you can see why Columbus first sailed southward from Spain to the Canary Islands searching for the east wind that blows from Europe to America. We can also see why he first sailed northward when he left the West Indies to find the west wind that would blow him back home to Spain.

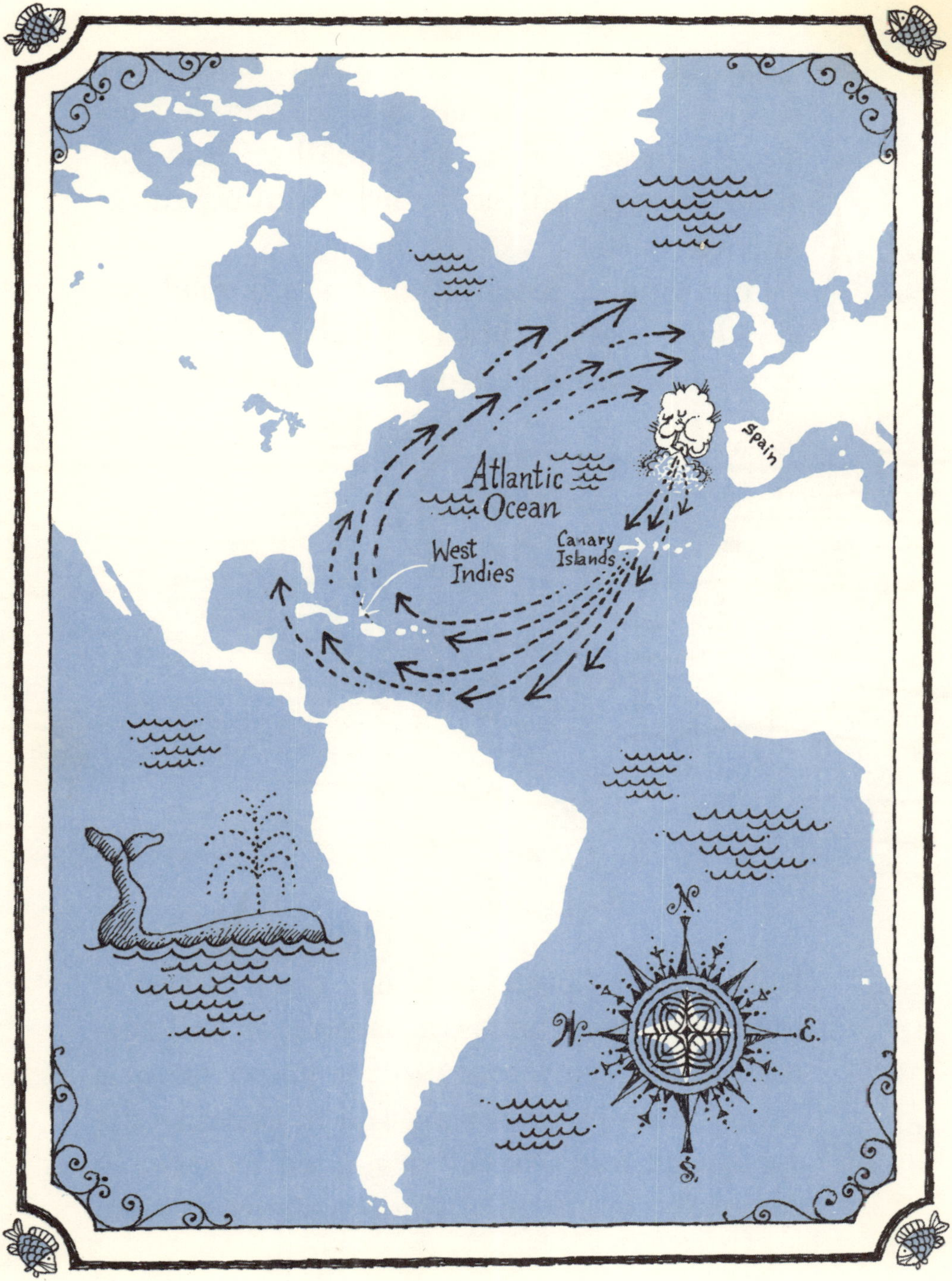
Atlantic Ocean
West Indies
Canary Islands
Spain
N
W
E
S

In these great rivers of air, there are patches of air that may be swirling as they move. These swirls are smaller than the main stream, but still may be hundreds of miles across.

One kind of swirling air mass is made up of heavy, cool air that is spiraling downward in a clockwise direction toward the ground.

It is called a "high" because it is a region of high *pressure;* that is, a barometer will register a higher pressure in the heavy air of a "high" than in the air around it. Because the air is cool and usually dry, good weather is generally found in a high-pressure region.

Another kind of swirling air mass is called a "low." It is made of lighter, warmer air that spirals upward in the atmosphere. Its pressure is lower than that of the air around it. A low usually contains rain and bad weather.

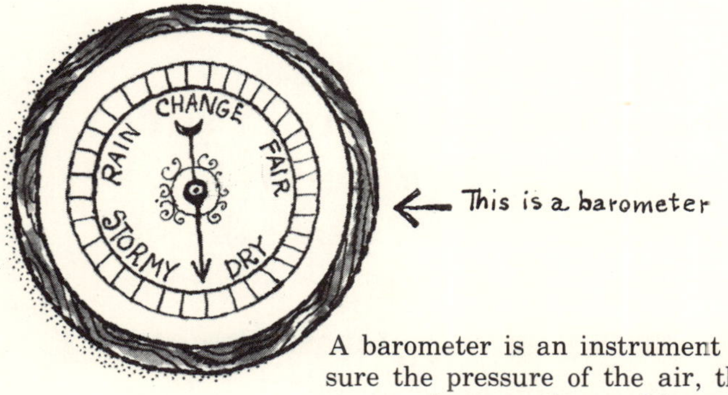

A barometer is an instrument used by weather forecasters to measure the pressure of the air, that is, the force with which the air presses down on the earth's surface.

Lows and highs are often shown on the weather maps we see on TV or in the newspaper. When you see a high, you know that it is an area of winds swirling slowly in a clockwise direction, while a low is an area of winds swirling in a counter-clockwise direction. (In the Southern Hemisphere these directions are reversed.) These slow whirlwinds move across the earth inside the big rivers of air. In North America, for example, lows and highs travel slowly across the continent in a generally west-to-east direction.

Near the equator, a low-pressure region spinning slowly over the ocean may speed up to become a tropical storm or even a dangerous hurricane. Such a storm might be thought of as a mass of air that has too much energy and needs to get rid of this energy by mixing with cooler air toward the pole. Hurricanes sometimes leave the ocean to smash into coastal areas with winds of 100 miles per hour or more.

Some local winds are caused by heating and cooling in that area. For example, during the day the air over the land warms and rises. A breeze blows from the cooler ocean toward the land.

During the night the land loses heat faster than the water, and becomes cooler than the ocean. The warmer air over the ocean rises and draws a cool breeze from the land toward the ocean.

Some places on the earth have winds that are very strong. Near the South Pole the wind often blows as fast as 200 miles per hour. The fastest surface wind ever measured was 225 miles per hour at the top of Mount Washington in New Hampshire.

In other places there may be very little wind. Even places where the wind is sometimes strong will at other times have no wind at all.

Although the wind is not very reliable as a source of power, it was for a very long time one of the few sources available. Windmills pumped water and ground grain whenever the wind blew. When there was no wind, the workmen sat by the mill and waited for a breeze. Sailing ships carried cargo and people across the oceans and around the world. When the wind died, the ship just drifted and its sails hung loose until the wind filled them again.

After the steam engine was invented, sailing ships were slowly replaced by ships driven by engines. Ships with engines could travel faster than sailing ships and could keep moving whether the wind was blowing or not. They could also travel a direct route from one port to another, while sailing ships had to travel by whichever route the wind allowed. The power to drive these ships came from coal or oil burned in the engine.

ROOM FOR RENT
ROOM FOR RENT

Windmills, too, were largely replaced by engines and electric motors. Many of the windmills in Holland do not pump water any more but have been made into museums or private homes.

In recent years, oil and coal have become scarcer and more expensive. We know that there will be a shortage of energy some day when the earth's supply of these fuels is all used up. People are beginning to think again about using wind power for some of our energy needs. New windmills to generate electricity are being built, often called wind turbines, and new kinds of sailing ships are being planned.

Although there is a tremendous amount of energy in the winds that blow all over the world, it is not easy to take this energy from the wind. A windmill that can capture a useful amount of energy, and convert it to electricity, must be large, complicated, and expensive.

The amount of energy that we can take from the wind depends upon how fast the wind is moving. If you hold your hand from the window of a car that is moving slowly, you will feel only a gentle wind and a small force on your hand.

If the car is moving fast, you will feel a strong breeze and a larger force on your hand.

A windmill needs a large force and must therefore be located in a place where the wind is strong. It must be mounted on a high tower above any trees or buildings that might block the wind.

The amount of energy we can take from the wind with a windmill also depends upon how large the windmill is. This small windmill has small blades and can supply enough electricity to light only a few lamps. It cannot supply enough electricity to operate all the equipment in a house, especially the major ones, such as refrigerator, water pump, and electric stove.

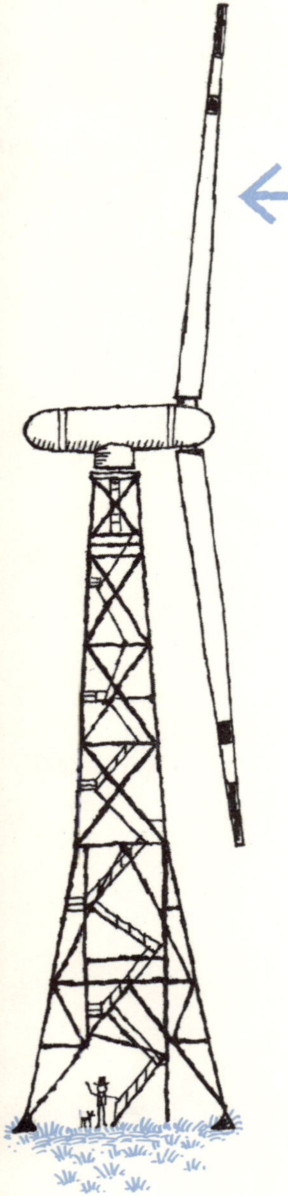

This large windmill built by NASA has blades 62 feet long and stands on a tower 100 feet tall. When there is a good breeze, this mill can generate enough electrical power for about 25 homes.

This very large windmill is as tall as a 12-story building and has blades 100 feet long. In a strong breeze, this mill will generate enough electrical power for 500 homes. Dozens of windmills like this one might be needed to generate enough power for a city. To replace one large electrical generating plant powered by nuclear energy, coal or oil, would require several hundred of these windmills.

Because windmills will generate electricity only when the wind is blowing, a city or a factory can not depend upon windmill power alone. A means for storing power, or equipment for generating electricity from some other source of energy will be needed when the wind is not blowing.

A city would need a lot of giant windmills like this to gather enough wind power for all its needs

Sails, too, must be very large to take a useful amount of energy from the wind. A sail as large as a handkerchief will make a model sailboat move. But a large sailing ship for carrying cargo and people must have tall masts and very large sails to gather enough energy from the wind to push the ship at a useful speed.

Wind power could supply a great deal of the energy we need for transportation and for our homes and factories. Wind power is free and non-polluting. It can never be used up because a new supply comes to the earth from the sun every day.

Perhaps in the not-too-distant future you will be able to travel across the ocean on a sailing passenger ship driven by wind power. You may soon see huge windmills whirling on mountaintops and in windy places by the ocean, making electricity from the earth's wind power.

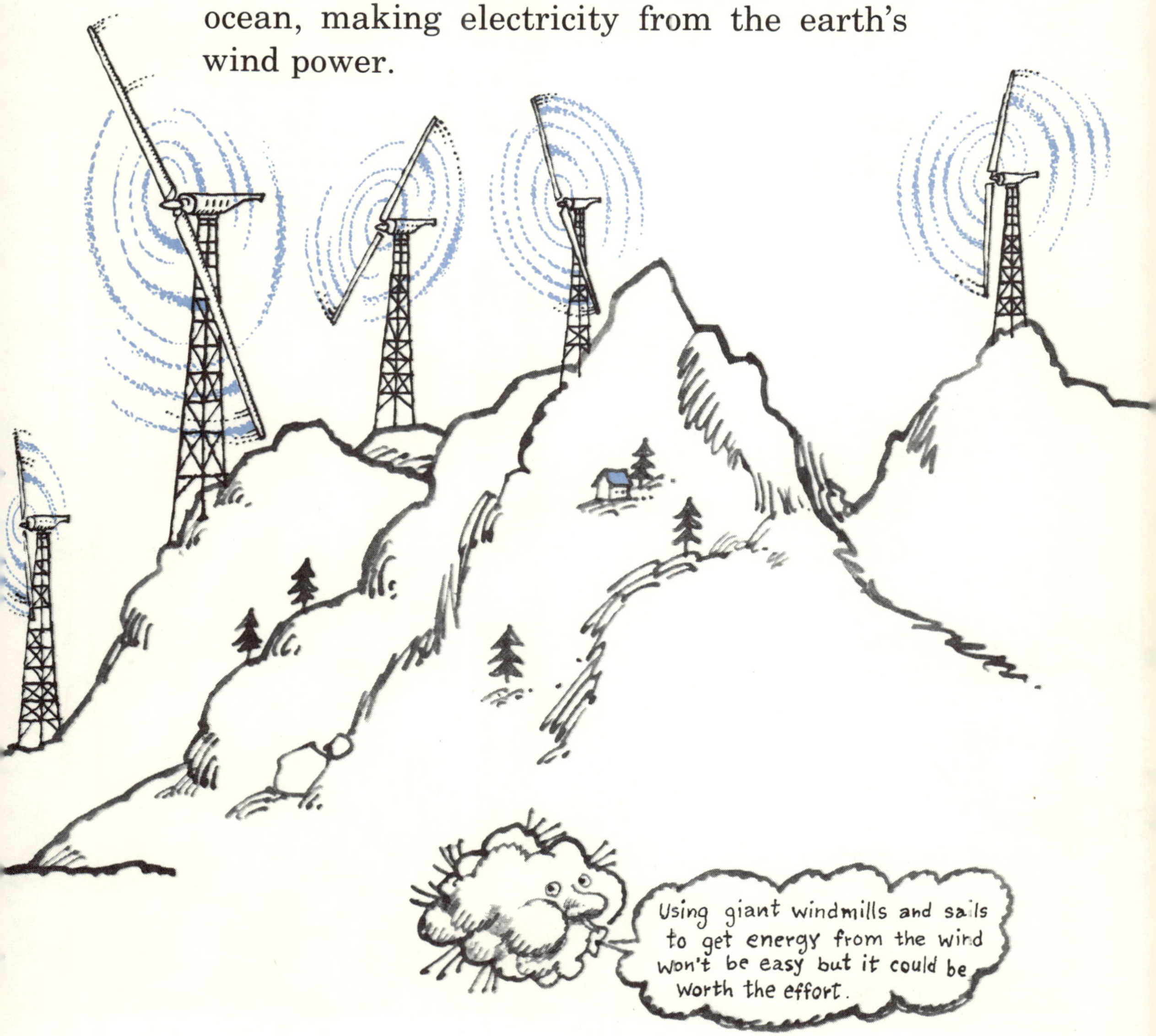

About the Author

Norman F. Smith received a degree in mechanical engineering from Purdue University and then went on to become an aerospace scientist with NASA. He retired in 1970 to devote his time to writing educational science material for young people. He is the author of, among other titles, *Space: What's Out There; Sun Power; Moonhopping;* and *If It Shines, Clangs and Bends, It's Metal.*

Mr. Smith lives in Grand Isle, Vermont.

About the Artist

Don Madden received his art training at the Philadelphia Museum College of Art and taught drawing there for several years after his graduation. His illustrations have appeared in *Good Housekeeping, Seventeen* and *Harper's Bazaar,* and he has illustrated dozens of children's books, including *Sand* by Sally Cartwright, and his own *Lemonade Serenade.*

Mr. Madden, his wife and their two children live in Ballston Spa, New York, in a renovated farmhouse.